The Mysterious, Magical Mangroves of St. John, U.S. Virgin Islands

by Caroline S. Rogers, PhD

Caroline S. Rogers

LEFT: Propagules on Red Mangroves are long, bean-like structures up to 18 inches long. CENTER: A propagule floats horizontally at the water's surface. RIGHT: After becoming vertical, propagules develop roots and begin to grow.

Introduction

Hurricane Hole within Virgin Islands Coral Reef National Monument, St. John, is peaceful and beautiful and the site of the least disturbed mangroves in the US Virgin Islands. No other mangrove areas in the Caribbean are known to have so many species of corals growing in them.

Red Mangrove trees grow in a narrow fringe along the shores of the bays in Hurricane Hole, Mary Creek, and elsewhere in St. John. Some White and Black Mangrove trees are mixed in with the Reds. The word "mangrove" is a general term that refers to Red, White and Black Mangroves as well as Buttonwood trees and others found growing with them.

Red Mangroves are the most spectacular. They have long "prop" roots that grow down from their trunks and anchor in the sand or mud, forming an elaborate network or labyrinth that provides shelter and hiding places for many species of fish and a settling surface for an amazing variety of other animals and plants, including corals, sponges, tunicates, and bryozoans.

Red, White and Black Mangroves are each in a separate family of plants, but all have special adaptations that make them more tolerant of salt in the water and the soil, allowing them to live where other plants cannot.

In each of these species, new plants develop directly from flowers and not from seeds. Seedlings with roots develop from "propagules." Propagules on Red Mangrove trees drop to the water where they float horizontally, sometimes for several months, until they come to a shallow place where they right themselves and put out roots. In this way mangroves can actually create or stabilize land, gradually extending the shoreline out to sea.

Most people would prefer to snorkel around Trunk Cay or on a coral reef than in the mangroves. In most mangrove-lined bays the water can be murky and turbid. It can be very spooky to swim in shady, dark areas where the branches of the mangroves extend out over the water, and it is difficult to see just what might be lurking there. However, in many places in Hurricane Hole it is easy to see into the roots, and

LEFT: *Princess Bay, Otter Creek, and Water Creek with Tortola in the background.*

ABOVE: *Brown seaweeds grow near the rocky shores between the mouths of the bays in Hurricane Hole.*

to see what is looking out at you. The water up near the roots is often clearer than farther from shore and in the middle of the bays, probably because sponges, oysters and other organisms are filtering the water continually.

In Florida and Puerto Rico, where mangroves are far more extensive than around St. John, more freshwater flows or runs off into the shallow water and fewer species of marine organisms grow on the prop roots. Especially heavy rains can lower the salinity (saltiness) of the seawater and kill animals growing on the roots. The lack of freshwater flowing into the bays in Hurricane Hole, except during occasional heavy rains, contributes to the remarkable diversity of animals found on the prop roots.

Although I knew that there were corals and colorful sponges growing in Hurricane Hole I did not do much snorkeling there until late in 2008 when I began to explore each of the bays more thoroughly. I was surprised to find an amazing diversity and number of corals growing on and near the prop roots of Red Mangroves in Princess Bay, Otter Creek, and Water Creek. (The word "Creek" does not refer to freshwater in this case). Although I know that corals have been seen in mangrove forests in Puerto Rico and Belize, I do not know of any other mangrove areas in the Caribbean with such a high number of coral species. I have seen 30 coral species in Hurricane Hole including a few individuals of species that are quite rare on the island's coral reefs, which have about 45 species. Borck Creek has very few corals, and only near the entrance to the bay.

Between the mouths of the bays, mangroves give way to rocky areas with higher wave energy where the trees cannot gain a foothold. Many corals grow just off these headlands.

Another welcome surprise, besides the diversity of corals, is that the corals growing in Hurricane Hole are generally healthier than most on the coral reefs. Some of these colonies are so large that they clearly survived the major bleaching episode and subsequent coral disease outbreak in 2005/2006 which had such devastating consequences for the reefs of St. John, the other US and British Virgin Islands and Puerto Rico.

In 2001, the day before he left office, President

Clinton established Virgin Islands Coral Reef National Monument, partly to protect the mangroves in Hurricane Hole which were recognized as a "nursery" for juvenile fish, Spiny Lobsters, and Queen Conchs. As these animals develop from larvae and juveniles into adults, many move from the shallow bays in Hurricane Hole to deeper waters offshore. These movements link the mangroves to seagrass beds, coral reefs, and other habitats inside and outside the monument.

The National Park Service and Friends of Virgin Islands National Park have installed a storm mooring system that eliminates the damage from the tying of boats to mangrove trees and reduces damage from anchoring. Moorings have been provided for day-use only, and no boats are permitted to stay overnight.

Regulations require boats to travel at slow speeds, and no water skiing or jet skiing is allowed.

It is tempting, and yes, fun, to zoom around in a fast boat in these typically calm bays, but the wake from even the smallest motor boats creates a large surge and stirs up the fine sediments on the bottom. These sediments rain down on all of the animals and plants growing on and near the prop roots and leave the water cloudy for over half an hour or more.

To protect the fragile and diverse animals and plants in the mangroves around St. John, and to avoid injuring yourself, it is important not to touch anything at all. Fireworms, fire coral, and fire sponges all have "fire" in their names for a reason. For example, fireworm spicules cause worse pain than a bee sting. It is easy to become almost mesmerized by the beauty on the prop roots—leading to smacking your head on an overhanging branch.

For hundreds of years, Hurricane Hole has been a haven for boats during storms. It is now a unique haven for an amazing variety of animals and plants, with its remarkable combination of mangroves, coral communities, and seagrass beds. The objective of this book is to help people discover an extraordinary ecosystem on St. John and to increase their awareness of the need to respect mangroves in general. Mangroves are even more endangered than coral reefs and tropical rain forests, with an estimated 35% of the world's mangroves already destroyed, many filled for aquaculture ponds or used as dumping grounds.

This Mountainous Star Coral, almost 3 feet across, has been growing here for several decades.

Boulder Brain Corals, such as the brown one in the center of this picture, are quite abundant in Hurricane Hole. They are major reef-builders because they reach a large size (over 6 feet across).

The Mysterious, Magical Mangroves of St. John, U.S. Virgin Islands

How This Book Is Organized

This book is divided into sections each of which focuses on a separate Phylum. All living organisms are classified into a hierarchical system that includes a Kingdom, Phylum, Class, Order, Family, Genus and Species. Each organism has a unique name consisting of its genus and species names.

For example, in Hurricane Hole you might see the Common Octopus, *Octopus vulgaris*. It is in the:

- **KINGDOM** Animalia
- **PHYLUM** Mollusca
- **CLASS** Cephalopoda
- **ORDER** Octopoda
- **FAMILY** Octopodidae
- **GENUS** Octopus
- **SPECIES** vulgaris

A Queen Angelfish (*Holacanthus ciliaris*) is in the **Kingdom** Animalia, **Phylum** Chordata (Chordates), **Class** Actinopterygii, **Order** Perciformes, **Family** Pomacanthidae, **Genus** *Holacanthus*, and **Species** *ciliaris*.

Vertebrates (animals with backbones) and invertebrates (animals that lack backbones) are two general groups, with vertebrates found only in the Phylum Chordata that includes humans, birds, fish, and sea turtles. Octopuses, squid, Queen Conchs, and feather duster worms are all invertebrates.

Around 100 Phyla (plural of Phylum) are currently recognized, with about 50 that include species found in the ocean.

***LEFT:** Brown plate-like Lettuce Corals and blue and red sponges grow on the prop roots. A Massive Starlet Coral is in the foreground.*

***OPPOSITE PAGE:** Some Boulder Brain Corals are a brilliant green and brown.*

PHYLUM: Cnidaria

CLASS: Anthozoa
SUBCLASS: Hexacorallia
ORDER: Scleractinia

CORALS

A remarkable number of coral species grow on and near the prop roots of the Red Mangrove trees in Hurricane Hole.

A true stony or "hard" coral is a group or "colony" of several polyps. Each one is similar to a sea anemone with tentacles around a mouth. They are connected at their bases and share a limestone support or "skeleton."

Almost all corals are colonial. An exception is Atlantic Mushroom Coral, a solitary coral that is a single polyp up to 6 inches across, within an individual limestone cup.

Flower Corals have a pretty shape with separate branches that end in one or two polyps. The polyps divide as the corals grow.

Grooved Brain Corals are some of the most common corals around St. John. They have ridges that separate valleys in which there are rows of connected polyps.

Corals are carnivores and have tentacles that can grab small animals, like zooplankton, out of the water column. They also function like plants because of the microscopic algae (zooxanthellae) that grow inside their tissues and provide nutrition and most of their brownish-greenish-yellowish color. These algae, like other plants, can photosynthesize, making food that the coral can use from sunlight, water and carbon dioxide.

In September 2005, seawater temperatures were the highest ever recorded (over 90º F) for this region of the Caribbean. Almost all of the corals around St. John became pale or white because the zooxanthellae disintegrated or were expelled. Some of the corals died because they were not able to get enough nutrition without these microscopic algae. However, when seawater temperatures began to go down, many corals began to recover.

Unfortunately, at this time there was an outbreak of a devastating disease called "white plague" that killed portions of or entire coral colonies. Over 60% of the coral around St. John died. Because corals grow very, very slowly, recovery could take decades. Recovery could be hindered or even prevented by future bleaching and disease events (as well as hurricanes) — all of which are predicted to happen more often because of climate change.

This coral, about 4 inches across, consists of just one limestone cup and one polyp. This coral is quite rare on the reefs around St. John, and usually found at least 30 feet deep. It is surprising to see it in very shallow water in Hurricane Hole.

Grooved Brain Corals.

A disease called white plague has killed a large portion of this Grooved Brain Coral. Fireworms are feeding on the coral tissue.

LEFT TO RIGHT: Bleaching (paling) occurs when the coral skeleton is visible through the coral tissue after loss of the symbiotic algae. This coral regained its normal color graduallly over a few months.

A bleached Grooved Brain Coral next to a Boulder Brain Coral with normal coloration.

CORAL BLEACHING IN 2005 AND 2010

No one was studying the corals in Hurricane Hole during the 2005 bleaching event, and we do not know how they were affected. However, some coral species in the area, particularly the Grooved Brain Corals, Finger Corals, and Lettuce Corals, bleached quite severely in the late summer and fall of 2010.

Research by the National Park Service revealed that most coral species on the reefs around St. John bleached at this time and a few died at least partially from disease. Because most corals are groups of polyps joined together, portions of them can die while the overall colony survives.

In the fall of 2005 seawater temperatures would have been especially high in the protected shallow waters in Hurricane Hole. The large corals that we see today were growing there during the bleaching event and must be more resistant than those that died on the coral reefs. High light intensity can also cause or contribute to coral bleaching. In October 2005 there were several days with exceptionally calm seas when more light would have reached the depths of the corals. The shade of the Red Mangrove trees in Hurricane Hole might have reduced the stress on some of the corals in these bays.

Many of the corals in Hurricane Hole are small and may have settled out of the water column as larvae and begun to grow since 2005. The presence of small "recruits" is a very good sign. It shows that conditions are good enough for new corals to settle and begin to grow.

TOP: *A Cactus Coral about 6 inches across.* **ABOVE LEFT:** *A Flower Coral recruit about 2 inches in diameter.* **ABOVE RIGHT:** *This Cactus Coral recruit is only an inch across.* **LEFT:** *This Flower Coral is about a foot across, larger than most colonies of this species.* **OPPOSITE PAGE:** *A juvenile Foureye Butterflyfish swims between Finger Corals and a Boulder Brain Coral.*

This Boulder Brain Coral in Water Creek is one of the largest corals in Hurricane Hole, almost 3 feet across.

Although quite different in color and shape, all of these corals are small Boulder Brain Corals.

PHYLUM: Cnidaria

CLASS: Hydrozoa

ABOVE & RIGHT: Fire Coral is often one of the first to bleach when sea water temperatures become very high. As you might imagine, Fire Coral gets its name from the intense stinging sensation you get if you brush against it. The stinging is from the barbs and poison of tiny capsules called "nematocysts" that fire if they are touched. Although all corals, anemones, and jellyfish have nematocysts, only some types of nematocysts cause severe pain.

PHYLUM: Cnidaria

CLASS: Anthozoa

Several species of Lettuce Coral grow abundantly in Hurricane Hole, many of them directly on the prop roots, and many on rocks close to the shoreline.

Rose Corals often are not attached to the bottom. They can take in water and turn themselves back upright if they are turned over. Few corals fluoresce, as shown here, in the daytime. Some research suggests that fluorescent pigments protect corals from damaging amounts of sunlight.

OPPOSITE PAGE: *Silversides stream past a large Mountainous Star Coral and several small colonies of Mustard Hill Coral, Boulder Brain Coral, and Massive Starlet Coral.*

Silversides swim over a patch reef of Finger Corals.

Finger Corals grow as isolated colonies on the mangrove prop roots as well as the bottom of the bays, but are also found in small patch reefs near the mouths of Otter and Water Creeks. This type of coral is one of only a few that extends its polyps during the day making the corals look fuzzy. When you go snorkeling or Scuba diving at night, you will see that many corals are using their extended tentacles to catch small crustaceans.

Pillar Coral is another species that usually has its tentacles extended during the daytime.

Golfball Corals are usually less than 3 inches across and grow abundantly on rocks in shallow water.

Cactus Corals are quite rare and are usually in deeper water on coral reefs. It is surprising to see them in shallow water in Hurricane Hole. Only 6 have been seen so far.

OPPOSITE PAGE: *Only a few Ivory Bush Corals have been found in Hurricane Hole. They are not common around St. John.*

Maze Corals are rare in Hurricane Hole.

A Flower Coral grows on a prop root.

LEFT: *Lettuce Corals, Maze Corals and Mustard Hill Corals grow in a few feet of water in the shade of the mangrove trees.*

PHYLUM: Cnidaria

CLASS: Anthozoa
SUBCLASS: Octocorallia

ABOVE: Silversides rush past sea plumes growing between Otter and Water Creeks.

TOP RIGHT & OPPOSITE PAGE: The stems of the Snowflake Coral extend from a prop root encrusted with sponges.

OCTOCORALS

Sea plumes, sea fans, sea rods, and sea whips (usually referred to as gorgonians) are in the Subclass Octocorallia. "Octo" means 8, and all of these have polyps with 8 tentacles. In contrast, members of the Subclass Hexacorallia ("Hexa" means 6), such as the corals and anemones, have polyps with 6 tentacles or multiples of 6. Unlike the stony (hard) corals, octocorals do not have rigid skeletons. Gorgonians are a defining feature of Caribbean and Western Atlantic coral reefs, and far more numerous on these reefs than those in the Indo-Pacific.

Considering their abundance, it is surprising that very few fish (including some filefishes, angelfishes, and parrotfishes) appear to eat them. The Foureye Butterflyfish nibbles on gorgonian polyps. Gorgonians provide shelter for fish and other animals.

Gorgonians are much more numerous on Caribbean reefs than on reefs in the Pacific.

A juvenile barracuda lurks near a sea plume with two Flamingo Tongue snails.

A sea rod on a prop root has its tentacles out to capture plankton that drift by.

Brown anemones with long tentacles grow next to a red Fire Sponge. A brittle star is visible near the top of the sponge.

PHYLUM: Cnidaria

CLASS: Anthozoa
ORDER: Actiniaria

ANEMONES

Anemones resemble flowers but are predators which eat microscopic plankton, small fish, and urchins. Much of the beauty in Hurricane Hole can be attributed to these colorful animals. Anemones can move around but usually stay in about the same place on roots or in crevices for several months.

The Giant Caribbean Anemone can be over a foot across.

Sun anemones are abundant between Otter and Water Creeks, forming mats on rocks or the bottom. They have short, stubby tentacles and can sting.

ABOVE & RIGHT: The tentacles of brown anemones growing on prop roots wave in the currents.

OPPOSITE PAGE: Spotted Cleaner Shrimp find shelter within anemone tentacles and pick parasites and dead tissue off of fish that pass by. If you wave your fingers gently near one of these shrimp it might come out to inspect your hand.

Beaded Anemones have a variety of different colors.

OPPOSITE PAGE: *The knobs on the tentacles of this Knobby Anemone have "nematocysts" (specialized structures with barbs) that can sting bare skin.*

OPPPOSITE PAGE &ABOVE: *Corkscrew Anemones have slender tentacles that they retract if disturbed.*

OTHER PHOTOS ON THIS PAGE: *Like corals, some anemones have symbiotic algae (zooxanthellae) in their tissues and will become pale or bleached when sea water temperatures rise and there is a loss of the zooxanthellae or their pigments.*

The Giant Caribbean Anemone has a body that is made up of a column with a mouth at the top, surrounded by tentacles. It can withdraw its tentacles completely.

A group of Pale Anemones resemble a bouquet of flowers at the end of a prop root.

OPPOSITE PAGE & ABOVE: *This Branching Anemone has long tentacles that are not branched and usually only extended at night, and shorter, stubby branched "pseudotentacles" that are evident during the day (shown here). It has a very powerful sting.*

RIGHT: *This Giant Caribbean Anemone is an unusual greenish-blue color.*

JELLYFISH

Except when there is an occasional "bloom" (a large increase in abundance) of jellyfish, you are not likely to encounter many of them around St. John. The two most common species are the Moon Jellies that have a pink or purple flower-like pattern on a round, transparent dome, and the Mangrove Upside Down Jellyfish that are common pulsating on the bottom in Hurricane Hole, especially Borck Creek.

The scientific name of the Upside Down Jellyfish, *Cassiopea*, is the name given to the constellation that looks like a "W" in the northern sky and represents the goddess in Greek mythology who was sentenced to be upside down forever for her vanity.

Like corals and anemones, these jellyfish have symbiotic algae (zooxanthellae).

Usually you see Mangrove Upside Down Jellyfish pulsating on the bottom but occasionally they rhythmically move through the water.

CLASS: Anthozoa
ORDER: Corallimorpharia

CORALLIMORPHS

Corallimorphs are similar to corals and anemones, but in a different Order (Corallimorpharia). They do not have limestone skeletons. They are not common around St. John.

ABOVE: *The disc-like corallimorph shown above has not been identified.*

LEFT: *This corallimorph (St. Thomas False Coral) has been referred to as Rhodactis sanctithomae because it was first described from St. Thomas in 1860. It has recently been reclassified and given the name Actinotryx sanctithomae.*

OPPOSITE PAGE: *French Grunts swim near corallimorphs, Red Rope Sponges, and corals.*

FISH

The colorful fish that dart back and forth in the waters around reefs, seagrass beds, and mangroves are sometimes compared to butterflies in a forest. Mangroves are critically important habitats for many young fish, including barracudas, snappers, and angelfishes. As these fish grow and mature, many of them, especially Schoolmasters, Foureye Butterflyfish, and Barracudas, move out to deeper water to seagrass beds and coral reefs. Some individuals remain as adults in the mangroves. Many fish in Hurricane Hole are not associated with the mangrove roots but rather with the seagrass beds, sandy plains, and rocky areas that are also found in these bays.

Fish are often grouped together based on what they eat. For example, some fish eat other fish (carnivores), while some eat mostly plants and algae (herbivores), and some eat a little of everything (omnivores). Much of what we know about the food that various fish in the Caribbean eat comes from studies by Dr. Jack Randall in the early 1960s around St. John. He examined the stomachs of over 5,500 fish from 212 species. He did not collect any fish from within the mangroves, but it is likely that the food preferences of the species are generally the same as when they are out in the seagrass beds or the reefs.

Some of the most conspicuous, and entertaining, fish in Hurricane Hole are the bizarrely shaped trunkfishes, like this Spotted Trunkfish, boxfishes, and porcupinefishes.

Grunts are the most abundant fish in the mangroves, schooling among the prop roots in groups of several hundred. Sometimes you can hear Grunts making noise underwater as they grind their teeth together.

ABOVE & LEFT: *Honeycomb Cowfish*

ABOVE: *Scrawled Cowfish*

Cowfish are similar to trunkfishes and in the same Family (Ostraciidae). Look for horns above their eyes!

At first it can be hard to tell the Spotted Trunkfish (ABOVE) *from the Smooth Trunkfish* (TOP LEFT) *but the Spotted Trunkfish usually has three pale areas on its sides. You might see the comical Smooth Trunkfish clowning around just above the bottom and blowing sand away to reveal tiny crustaceans to eat. The larger and rarer Buffalo Trunkfish* (LEFT) *is the third trunkfish species found around St. John. You are most likely to see it swimming in seagrass beds.*

ABOVE & FAR RIGHT: *Porcupinefishes eat sea urchins, including Long-spined Black Urchins, Reef Urchins, and Slate-Pencil Urchins, as well as snails ("gastropods") and crabs. The same Porcupinefish may be seen over and over, month after month, hiding in the prop roots in the same location.*

RIGHT: Sharpnose Puffers, less than 4 inches long, are omnivores, feeding on plants, sponges, and worms.

A Queen Angelfish peers out from its hiding place behind roots covered in sponges and algae.

Angelfishes feed almost exclusively on sponges, in spite of the sponges' glass-like skeletal needles called spicules. Queen Angelfish, both adults and young juveniles with vertical blue bars, feed up in the roots more often than French or Gray Angelfish. Perhaps there are different sponges growing on the prop roots than in slightly deeper water, and these are the ones that Queen Angels prefer. Rock Beauties (next page) are rare but sometimes are seen feeding in the roots as well.

ABOVE: *An intermediate French Angel near roots covered with sponges.*
BELOW: *A juvenile Gray Angel nips at a sponge.*

A Rock Beauty feeds in the roots.

PHYLUM: Chordata

Colorful and active little butterflyfishes swim quickly among the prop roots and corals, using their elongate snouts to pick small animals and sometimes polyps from crevices in corals. Juveniles often have different markings and color patterns than adults. Very young Foureye Butterflyfish have an extra spot near their tails. Juveniles are often alone but adults are usually in pairs.

ABOVE: *A juvenile Foureye Butterflyfish.*
LEFT: *An adult Foureye Butterflyfish.*
BELOW: *A juvenile Banded Butterflyfish.*

PHYLUM: Chordata

LEFT, ABOVE & OPPOSITE PAGE: *Gray Snappers, also called Mangrove Snappers, live as full-sized adults in the mangroves. Schools of these large fish can be quite startling when they suddenly appear in the labyrinth of prop roots.*

A Queen Angel and Gray Snapper.

Barracudas of all sizes live in the mangroves and are very abundant in Hurricane Hole. They range in size from tiny, nearly transparent juveniles about 1 inch long to a few large ones up to 3 feet long.

Tiny "cudas" less than an inch long and almost transparent hover just beneath the water's surface often showing a bit of "attitude" with their body slanting upwards.

A blizzard of silversides darts amongst the prop roots over a lush seagrass bed.

Sergeant Majors are some of the most common fish on the coral reefs around St. John, and they are also seen in Hurricane Hole most often near the rocky, coral areas. You may see them aggressively guarding their purple eggs which they deposit on sides of rocks.

OPPOSITE PAGE: *A school of Horse-eye Jacks above the sandy bottom in Otter Creek.*

A Sand Diver, one of the most common species of lizardfishes in the Caribbean.

If you see a lizardfish, it will most likely be lying on the bottom. These fish will remain motionless for long periods of time and then streak up into the water column to snatch a fish. It is rare to witness predation, but once I saw a lizardfish with lightning speed launch itself towards the surface, grab a Yellowfin Mojarra (a fish about 4 inches long) in its mouth, and then go back to lie in about the same place on the bottom—once again barely moving but this time with the head and tail of its prey sticking out of either side of its wide mouth.

Squirrelfish and other related species that are nocturnal have big eyes that allow them to hunt for shrimp and other crustaceans at night.

A Squirrelfish near a Beaugregory damselfish.

OPPOSITE PAGE: *A school of Blue Runners.*

A Tarpon, close to 4 feet long, instills fear in a school of tiny silversides.

The diversity of habitats in Hurricane Hole, with its coral communities, seagrass beds and mangroves, supports a high number of fish species.

TOP LEFT: A Peacock Flounder.

ABOVE: A Blue Tang.

LEFT: An unusual hybrid between a rare Indigo Hamlet and a Barred Hamlet.

Stingrays swim just above the sandy bottom, stopping to feed on mollusks and worms buried in the sand. Above their tail, near where it joins the rest of the body, is a poisonous serrated spine that can cause a severe wound. The danger comes from accidentally stepping on one of these fish and not from their aggressive behavior.

Remoras attach to sharks and large sea turtles, so whenever you see one you have to wonder what it fell off of! They eat parasites on their host and scraps left from whatever the host is feeding on. I have only seen a few sharks (a Nurse Shark and some juvenile Blacktips) in Hurricane Hole, but juvenile Lemon, Nurse, and Blacktip Sharks are seen in the mangroves in Mary Creek and Coral Bay Harbor (behind Skinny Legs Restaurant). There is a story of a large Bull Shark in Water Creek, but that was many years ago.

Permits are some of the largest fish that you might see cruising over the sandy and rocky areas in Hurricane Hole. They eat mostly clams and snails.

THE SARGASSUM FROGFISH

One day when I was snorkeling in Princess Bay, something caught my attention in the seaweed growing on a mangrove prop root—it appeared to be a sort of yellow blob. When I looked more closely I noticed the blob had a tail, and I realized that it was a fish, less than 2 inches long, clinging to the seaweed.

Later, I checked a fish identification book and decided the fish was probably a Sargassum Frogfish that usually lives in floating mats of brown seaweed (*Sargassum*) in the open ocean. This seaweed gives its name to the Sargasso Sea, an area in the middle of the North Atlantic surrounded by ocean currents. The seaweed floats near the surface and provides a home for a diverse community of fish, sea turtles, and mollusks.

Frogfishes float into the bays on Sargassum seaweed that wraps around the prop roots.

I sent the picture that I had taken of the fish to Dr. Jack Randall, an internationally renowned ichthyologist (fish expert), who lived on St. John from 1959 to 1961, doing research that was the basis for several now classic papers on fish and conchs. He sent my picture along to Dr. Ted Pietsch, an expert on frogfishes, at the University of Washington, who confirmed that it was indeed a Sargassum Frogfish (*Histrio histrio*).

I realized that this single fish could just indicate a chance occurrence, but I began to search for others in Hurricane Hole. Over the course of the next few months, I was thrilled to discover over 25 more, ranging in size from about 1 inch to over 4 inches long.

These fascinating fish don't look much like frogs, but they look more like frogs than fish. They have prehensile paired fins allowing them to move slowly over the algae they are hiding in. The pectoral fins have elbow-like joints. They are not very fast swimmers and rely on camouflage, blending in remarkably with all their frills and yellowish-brown color patterns.

I have found individual Sargassum Frogfish in the same clump of algae growing on prop roots over several weeks. However, I do not think that they survive for more than a few months, most likely because they are eaten by snappers or other predatory fish. Given their bizarre appearance, one would think fish might hesitate to eat them! The Sargassum Frogfish are carnivorous, eating crustaceans and other fish—even other Sargassum Frogfish! Dr. Pietsch told me that he once found 14 of them inside one large individual.

The discovery of this fish in Hurricane Hole raises many questions. Do they all drift in within clumps or mats of *Sargassum* seaweed? How long do they survive in the mangroves? Do any reproduce in here? Where do the floating clumps of seaweed come from—are they broken off of extensive mats far out at sea? What do they tell us about water circulation and currents within Hurricane Hole?

A few months after discovering the Frogfishes I came across three more animals that are associated with floating *Sargassum* seaweed—an Opossum Pipefish, a sea hare and a Sargassum Crab. The pipefish was swimming near a clump of algae growing on a prop root. Amazingly, this fish requires many different habitats to

complete its life cycle. Adults move into freshwater ponds to breed and the young move or are carried out to brackish and salt water where they mature, most likely within floating mats of *Sargassum* or other types of seaweed. The fish is very rare and I would be surprised to see any more of them.

Why did I notice the inconspicuous but unusual Sargassum Frogfish? All of us have had the experience of realizing that something we see is unusual, without consciously being aware at first as to why we perceive it as different. I recall the first time I saw a brilliant yellow and black Hooded Warbler on the railing of a deck at the house that I was living in at the time in Catherineberg. Those of us who live on St. John are very familiar with the Bananaquit or Sugar Bird that has similar colors. Something about the behavior of this bird or its color pattern made me stop and take a closer look. This chance observation of a species that I had never seen before led me to search for more warblers and I eventually saw several other species that migrate to or pass through St. John every winter. Seeing the unusal Sargassum Frogfish and the visiting warblers increased my appreciation of the ecological importance of St. John.

Like the Sargassum Frogfish, this Sargassum Swimming Crab and unidentified sea hare probably floated into Hurricane Hole on mats of the brown seaweed.

PHYLUM: Chordata

One day I saw a Hawksbill holding this rock in its mouth. After it dropped the rock, I saw it had taken bites out of the encrusting Chicken Liver Sponge.

Both Hawksbill and Green Sea Turtles live in Hurricane Hole, although it is fairly uncommon to see them. The ones that you come across usually swim away rapidly, unlike in Maho and Francis Bays where they are more accustomed to snorkelers. Some Hawksbills come up to nest on the little beach in front of Zootenvaal near Borck Creek.

Hawksbills feed almost exclusively on sponges, although they also consume zoanthids and occasionally algae. One of their favorite foods is the Chicken Liver Sponge that encrusts rocks in all of the bays in Hurricane Hole.

One day while snorkeling in Mary Creek, I came across a low rock wall that seemed to be man-made. I later learned that this was a turtle pen, or kraal, dating back to the late 1700s. These pens enclosed the turtles but had large enough openings to allow water to rush in and circulate. The turtles were fed until it was time for them to become dinner for someone. There are records of other turtle kraals around St. John, including one in Hurricane Hole, although its location is not known.

THIS PAGE & OPPOSITE PAGE:
Hawksbill Turtles swimming and resting on the bottom in Water and Otter Creeks.

PHYLUM: Chordata

SPONGES

Sponges are in the Phylum Porifera and are the most primitive group of animals that have many cells. Some species tend to be found more in mangrove areas, and others on coral reefs. About 60 species of sponges grow in Hurricane Hole, and many of the species are more often associated with reefs than mangroves. A surprising variety of animals use sponges for food and shelter—fish, sea stars, brittle stars, shrimp and lobsters. Much of the color in the mangroves comes from sponges which are just about every color imaginable and have a wide variety of shapes reminiscent of vases, barrels, and bushes. Most sponges have spicules (skeletal needles) made of silica although some have spicules made of calcium carbonate.

LEFT: *It is best to not touch anything at all in the mangroves, especially these red Fire Sponges that can cause a severe burning pain, lasting several days.*

Two of the most common sponges in Hurricane Hole are the Red Rope Sponge (ABOVE) and the Chicken Liver Sponge (RIGHT).

Sponges contribute much of the color seen in the mangroves and on the bottom in Hurricane Hole.

The circular opening (siphon) of a reddish ascidian reveals its presence amongst sponges, anemones, and algae.

ASCIDIANS

Ascidians, also called sea squirts, are bizarre animals. There is nothing like them on land. They look like invertebrates but they are classified in their own Subphylum (Tunicata) in the Phylum that includes all vertebrates (Chordata) because in their early life cycle they have characteristics common to vertebrates, including a primitive backbone. They are actually more closely related to us than most of us would like to acknowledge. For example, they are more closely related to us than the far more appealing squid to which some of us might feel more of a bond.

Although ascidians are quite common on reefs, seagrass beds and mangroves, and some are brightly colored, most of us don't notice them at all. Encrusting ones especially can be difficult to distinguish from sponges, but if you wave your hand over their siphons (openings) they will usually react by closing them. Sponges do not react (visibly) to anything at all!

PHYLUM: Chordata

SUBPHYLUM: Tunicata

A TUNICATE THAT FIGHTS CANCER!

I have only seen one colony or cluster of the especially pretty Orange Mangrove Tunicate (*Ecteinascidia turbinata*) growing on a prop root in Hurricane Hole although it is abundant in some other mangrove areas in the Caribbean. It resembles the Blue Bell Tunicates that are common near Trunk Cay. Each individual in a colony is an inch long or less.

Decades ago researchers discovered that extracts of this tunicate had powerful anti-cancer activity, and the active compound was characterized and named ecteinascidin. Very large amounts of the tunicate were required to get even small amounts of the anti-tumor material. However, recently a compound derived from ecteinascidin has been synthesized, and it is showing great promise in fighting several different types of cancer.

PHYLUM: Chordata

SUBPHYLUM: Tunicata

Ascidians have openings called siphons. They feed by pumping water into one siphon, through a gill net, filtering out their food, and discharging the water through another siphon. If you stir up sediments by kicking with your fins or by driving a boat too fast through the bays in Hurricane Hole the particles can clog this filtering mechanism, leading to the death of the organism.

Some tunicates, like the velvety Black Tunicate, are solitary but others grow in colonies or in mats.

ABOVE: *Black Tunicates are some of the most conspicuous ones growing on the prop roots.*

RIGHT: *An encrusting, colonial tunicate.*

ECHINODERMS

The Phylum Echinodermata includes sea stars, sea urchins, sea cucumbers and brittle stars. It is unusual in having only marine representatives (no freshwater or terrestrial species).

Sea stars (also called starfish) eat sponges and leave the skeletons behind. They also eat algae in the sand and occasionally tunicates. In Belize, juvenile Cushion Sea Stars (up to 10 inches across) are found in mangrove lagoons but here many of the ones we see are adults.

The Long-spined Black Urchin has black (and sometimes white) spines that can be over a foot long. It can be very painful if you accidentally step on or put your hand down on one of these urchins. It is good not to touch anything at all while snorkeling in coral reef and mangrove areas!!! The spines are very sharp and have toxins. They are difficult to extract because they have barbs on them.

Most of the Long-spined Black Urchins died throughout the Caribbean in the early 1980s and many people rejoiced. However, these urchins help the reefs by grazing on algae, opening up space for coral larvae to settle and grow, and it is good that they are now becoming more numerous.

ABOVE: Long-spined Black Urchins can have black or, rarely, white spines.

OPPOSITE PAGE: A Cushion Sea Star.

ABOVE: *A Variegated Urchin.*

RIGHT: *Brittle stars usually hide during the day, emerging at night to feed on particles of organic matter in the sand and small crustaceans and worms.*

BELOW: *A Slate-pencil Urchin between Lettuce and Finger Corals.*

Sometimes Cushion Sea Stars climb up the mangrove prop roots to feed on sponges.

Some people eat the gonads inside this sea urchin, called a West Indian Sea Egg. The white shell that remains when the urchin dies is called a "test."

RIGHT & FAR RIGHT: Red Heart Urchins are usually buried in the sand during the day.

BELOW: The test or skeleton of a West Indian Sea Egg.

LEFT: Sea cucumbers are somewhat like earthworms in that they get their nutrition from algae and debris in the sediment. These elongated echinoderms have suction-like tube feet which propel them along the bottom.

Bright red Reef Urchins graze on algae growing on prop roots and boulders.

Reef Urchins and Flat Tree Oysters.

PHYLUM: Arthropoda

CLASS: Crustacea

The largest crab in the mangroves is the Giant Hermit crab that can be up to a foot long. This crab may have moved into this Queen Conch shell after an octopus pulled the conch out, or it may have killed the conch first and then moved into its shell.

CRABS

Crabs are very important in the diet of many mangrove animals, including fishes. Except for the Yellow Arrowline Crab, they are not very conspicuous. Some hide in the Giant Caribbean Anemone.

ABOVE: A Banded Clinging Crab hides near a withdrawn Giant Caribbean Anemone.

BELOW: A Yellow Arrowline Crab on a Fire Sponge.

Above: A Mangrove Crab on a Red Mangrove prop root.

ABOVE & BELOW: Spotted Cleaner Shrimp often live on Giant Anemones. If you wiggle your fingers in front of one it might come out to clean them!

Banded Coral Shrimp (ABOVE & OPPOSITE PAGE) clean parasites off of fish. Blue eggs are visible under the female (TOP LEFT).

LEFT: *Adult Spiny Lobsters hide under rocks and in the roots, and juveniles just a few inches long live in algae on prop roots and boat mooring lines.*

ABOVE: *This small lobster in Mary Creek was in a den near a pretty green zoanthid. Zoanthids are related to anemones and corals.*

OPPOSITE PAGE: *An unidentified crab, about 4 inches across, hides under an anemone.*

MOLLUSKS

The Phylum Mollusca has an amazing diversity of some of the most beautiful and interesting animals in Hurricane Hole, including Queen Conchs, sea slugs, Flamingo Tongues, squid and octopuses. Many fish eat mollusks, especially porcupinefishes, Balloonfish, and Spotted Eagle Rays.

The lovely geometric patterns of the Flamingo Tongue snail are not on its shell but rather on the tissue (called "the mantle") that secretes the shell and extends around it. All mollusks have this specialized tissue.

Squid use their tentacles to seize small fish, other mollusks, and crustaceans and pull them to their mouths. They use a parrot-like beak to cut up their prey.

Octopus are some of the most fascinating animals in Hurricane Hole. Usually, they hide in their dens during the day. They eat mollusks and leave their shells outside their dens. When you see a pile of empty snail or clam shells, look closely to see if you see an octopus, or part of one. They can change colors and textures.

There is some reason to believe that squid have a sense of humor.

Helmet Conchs can rise up over sea urchins and eat them.

Juvenile Queen Conchs stay buried in the sand until they are large enough to have a better chance of not being eaten! Nurse Sharks, crabs, and other animals eat small conchs, but the hard shells of adult conchs are difficult for most animals to get through—an exception is the Spotted Eagle Ray.

If you approach very slowly, you might see a Queen Conch sweeping its proboscis across the bottom as it feeds on algae.

ABOVE: *True Tulips are voracious predators on small conchs. The living animal is bright orange.*

RIGHT & ABOVE RIGHT: *Triton's Trumpets, the largest gastropods around St. John, are very rare.*

ABOVE: *The Rough Fileclam reveals its bright red mantle.*

BELOW: *Amber Penshells are found on sandy and hard bottoms.*

Flat Tree Oysters are bivalves that grow in great abundance on most of the Red Mangrove prop roots, within a few inches of the water surface. If you are hungry enough, you can eat these but there isn't much to them. Surprisingly, I have never seen anything eating these, except for one time when I saw an Apple Murex (above) which seemed to be eating a Tree Oyster. Some starfish eat oysters and clams, but the Cushion Sea Star that is common in Hurricane Hole eats sponges.

Chitons, often on rocks along the shoreline, have 8 overlapping, calcareous plates. You will get bored if you wait to see them move.

ABOVE: *Sea slugs and sea hares are occasionally seen feeding on algae on the prop roots.*

ABOVE & BELOW: *Sea slugs can be very colorful. Some, like these Lettuce Sea Slugs, are "solar-powered" because they have symbiotic chloroplasts in their bodies. They ingest these when they feed on algae.*

ALGAE

Algae grow just about everywhere on coral reefs, in mangroves, and in seagrass beds as well as on boat bottoms and anything that ends up in the water. Although essential to marine ecosystems, they are often overlooked because many are small and not too colorful. They provide food for many different kinds of animals and oxygenate the waters. Major groups of algae are referred to informally as red, green and brown algae. Here are a few of the more conspicuous species in the mangroves.

ABOVE: *A brown seaweed grows attached to rocks on the bottom.*

OPPOSITE PAGE: *A feather-like species of the genus Caulerpa grows near a Giant Caribbean Anemone.*

ABOVE: *Segments of this species and others in the genus Halimeda are important constituents of beach sand.*

ABOVE CENTER: *A species of green algae.*

ABOVE RIGHT: *Two different species of "Shaving Brush" algae grow surrounded by Manatee Grass (cylindrical blades) and Turtle Grass (flattened blades).*

RIGHT: *A species of red algae.*

ABOVE & BELOW: *Sea Pearls are macroscopic single cells (up to 2 inches across) that have a reflective surface and stand out, especially to beginner snorkelers.*

LEFT: *Unidentified red algae growing on a prop root.*

PHYLUM: Annelida

CLASS: Polychaeta

WORMS

We don't usually think of worms as attractive, but some of the worms living in the mangroves are very beautiful. It is likely that your attitude about worms will change once you become acquainted with the ones in the mangroves.

All of the conspicuous worms (many are hidden in the sand) are Polychaetes (in the Class Polychaeta).

Some polychaetes have colorful tentacles that they use to strain small food items from the water.

ABOVE: *An unidentied feather duster worm surrounded by algae and sponges.*

RIGHT: *Split-Crown Feather Dusters grow out of a Mountainous Star Coral.*

OPPOSITE PAGE: *A Magnificent Feather Duster Worm.*

PHYLUM: Annelida

CLASS: Polychaeta

Fireworms are very abundant in the mangroves, crawling along the bottom. Sometimes they feed on live coral including Fire Coral. Most are about 4 inches long, but they can attain a length of 1 foot.

It isn't clear if these worms only eat tissue from corals that are already weakened, for example by bleaching or disease, or if they will attack healthy colonies.

Fireworm feeding (predation) certainly exacerbates the loss of tissue from disease and sometimes leads to death of the entire coral colony. Fireworms may transfer disease from one coral to another.

If you brush up against a fireworm, touching the bristles along the margins of its body, the pain may be worse than a bee sting and can last several hours.

PHYLUM: Annelida

CLASS: Polychaeta

ABOVE: *Christmas Tree Worms have a small plate like a door that snaps shut quickly after the worm withdraws into a tube. Small eyespots sense changes in light if there is motion near the worm, and it pulls into its protected tube.*

TOP LEFT: *Split-Crown Feather Duster.*

LEFT: *Spaghetti Worms are well-named. They have long, white filaments that they extend over everything nearby as they search for small food particles. They transport the food back to their mouths (usually not visible) with cilia.*

RIGHT: *Not many fish eat fireworms but many do eat other polychaetes like this Feather Duster.*

BIRDS

The mangrove forest also provides a home or resting spot for many birds, including the Yellow Warbler—the only warbler that lives year round on St. John—and migratory warblers such as the Blackpoll, which passes through for a few days each year. In spite of their name, Mangrove Cuckoos spend more time in dry forests on the island than in the mangroves. Green Herons and Night-Herons eat fish and crustaceans. American Oystercatchers poke along the tide line searching for clams among the rocks.

LEFT: *Green Heron.*

ABOVE CENTER: *Yellow-crowned Night-Heron.*

ABOVE TOP RIGHT: *American Oystercatcher.*

ABOVE BOTTOM RIGHT: *Blackpoll Warbler.*

Brown Pelicans catch fish in their large bills, drain the water out of their pouches, and then swallow the fish whole.

CONCLUSION

We still have much to learn about the mangroves of St. John. There is a nearly complete list of the corals and the fishes, but not of the thousands of other plants and animals living in these fascinating habitats. St. John's mangroves are not as extensive, and the individual mangrove trees are not as large, as in many other places in the Caribbean. However, the high diversity of stony corals that have become established appears to be without parallel in the Caribbean. Further research is needed to help us get a better understanding of how these corals and other organisms in Hurricane Hole are connected to the true coral reefs around the island. Currents carry larvae from corals, fish and other organisms into the mangroves but also may transport coral larvae from the mangroves out to the reefs that have been so degraded by bleaching and disease, helping them to recover.

These mangroves are noteworthy not only for their beauty and their possible connections to other ecosystems but also for what they might be able to tell us about the effects of climate change on corals. Worldwide, destruction of mangrove forests has exceeded that of coral reefs. Both of these marine ecosystems face new challenges from global climate change, with rising sea level a concern for mangroves, and higher seawater temperatures and ocean acidification the greatest threats to coral reefs.

Scientists believe that local management efforts, including steps to reduce sedimentation, cutting of coastal vegetation, pollution, and physical damage from anchors and boat groundings, can increase the chances that coral reefs and associated mangroves and seagrass beds will persist in spite of climate change. If people understand the value of these ecosystems, they will be more likely to do what they can to protect them. It is my hope that this book has enhanced your appreciation for the unique and mysterious mangroves of St. John.

ABOVE: *Bryozoans are small, colonial animals made up of "zoids." The only conspicuous bryozoan in Hurricane Hole is the pretty Tubular-Horn Bryozoan.*

Index

Index